TUNNELING OPERATIONS AND EQUIPMENT

Proceedings of the session sponsored by the Construction Division of the American Society of Civil Engineers in conjunction with the ASCE Convention in Detroit, Michigan

October 22, 1985

Edited by D. D. Brennan

Published by the
American Society of Civil Engineers
345 East 47th Street
New York, New York 10017-2398

Library of Congress Catalog Card No.: 85-72339
ISBN 0-87262-483-8
Manufactured in the United States of America.

PREFACE

The papers included in the Proceedings describe four diverse tunneling projects of varying diameters, design criteria, ground conditions and construction techniques undertaken at locations throughout the United States.

The unique method of multiple drift construction of the Mount Baker Ridge Tunnel in Seattle, Washington, presented a challenge to the contractor never before undertaken.

The Flint, Michigan, sewer tunnel constructed over a period of three years was bored through soil conditions that varied from rock to running ground and utilized, by necessity, two different types of tunnel boring machines.

Large diameter, continuous, soft ground and hard rock tunnel boring machines were used for the construction of the Greenbelt Tunnels in Washington, D.C. and the deep tunnel TARP Project in Chicago, Illinois. The presentation and favorable acceptance of an extensive Value Engineering Change Proposal in Washington, D.C. and the new generation of tunnel boring machines in Chicago present a study in sound construction management and organization.

Throughout the Proceedings the authors describe the experience and construction approach taken by several different contractors for the successful completion of these varied and considerably difficult underground projects.

Each of the papers included in the Proceedings has been accepted for publication by the Proceedings Editor and has also received one positive peer review. All papers are eligible for discussion in the *Journal of Construction Engineering and Management.* All papers are eligible for ASCE awards.

CONTENTS

Construction of Mt. Baker Ridge Tunnel

John F. MacDonald, M. ASCE*

Mt. Baker Ridge Tunnel is a large diameter soft ground highway tunnel constructed through sensitive over-consolidated silts and clays. Its large diameter necessitates the construction of the permanent liner, consisting of concrete filled perimeter drifts, before beginning the excavation of the tunnel interior. This article describes the construction methods and equipment used to excavate and concrete the perimeter drifts.

Introduction

The Mt. Baker Ridge Tunnel is located within the metropolitan limits of Seattle, Washington. It is part of a general upgrading of the western most seven miles of Interstate 90. Initial studies, conducted in the 1960's and 1970's by Washington State DOT and their consultants, looked at several options for penetrating the 55 meter high ridgeline. These included open cutting the hill, constructing a series of smaller tunnels and constructing a single, multi-level, tunnel. Community and environmental concerns finally dictated the use of the single tunnel option. In its ultimate configuration, this large diameter (25 meters) tunnel will carry three westbound lanes and two reversible lanes of Interstate 90. The material of the ridge through which the tunnel is constructed, combined with the large diameter of the tunnel, led to its unique method of construction.

Construction Criteria

The general criteria for constructing the tunnel was outlined in the Contract Plans and Specifications. A Design Summary Report, made a part of the contract documents summarized the design considerations and rationale. The tunnel is constructed by the "stacked drift" method. This method consists of excavating a series of smaller tunnels, or drifts sequentially around the perimeter of the main tunnel and backfilling them with concrete. This concrete structure is then the final support when the interior of the large tunnel is excavated. (See Figure 1). This procedure permits the construction of the needed large diameter tunnel using equipment and construction methods adapted from more conventional sized tunnels. The Contract Documents call for the excavation and concreting of the lowest, or invert drift first. Once this is accomplished, adjacent drifts can be constructed and backfilled with concrete. This sequence is followed until the final or crown drift at the top of the tunnel had been constructed.

*Construction Manager, Guy F. Atkinson Construction Company, P. O. Box 1158, Mercer Island, Washington 98040.

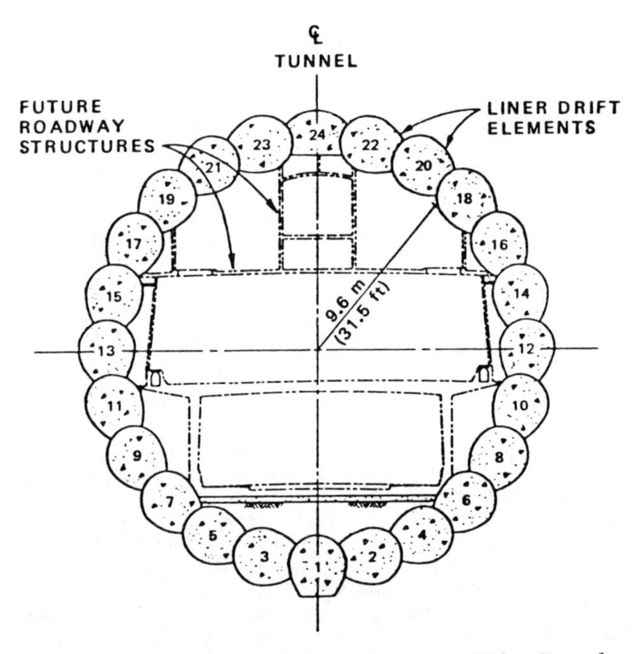

Figure 1. Cross Section of Mt. Baker Ridge Tunnel

Design considerations placed constraints on the number and size of the perimeter drifts. A minimum of twenty-four (24) equal sized drifts were specified in order to give the final liner the desired flexibility. In order to carry the anticipated ring thrust, a minimum contact width between adjacent drifts of 1.5 meters was specified. Joints between adjacent drifts were to be oriented radially to the larger tunnel. Working within these criteria, the contractor was free to choose the shape and number of the perimeter drifts to be constructed.

The material through which the tunnel is constructed loses much of its strength once it has been disturbed. For this reason the Contract Specifications limited the allowable subsurface settlement. During the construction of an individial drift the allowable settlement, as measured one meter above the drift, was limited to 25 mm. The State DOT monitored the amount of subsurface settlement via a series of instrumentation stations positioned over the tunnel.

Bids for the construction of the Mt. Baker Ridge Tunnel were opened on September 23, 1983. Guy F. Atkinson Construction Company's bid of $38,272,282 was $180,000 lower than the next bidder and 40 million less than the Engineer's Estimate. Construction began in December, 1983 with drift excavation commencing in September, 1984.

Figure 2. Shield Showing The Addition Of The Hood For Drift 2

The perimeter drift configuration chosen by the contractor consists of twenty-three (23) horseshoe shaped drifts and one hour glass shaped closure drift at the crown. The modified horseshoe shaped drifts were excavated using a backhoe type hoe mounted inside a shield, while the final drift was excavated by hand.

Two shields, fabricated especially for the project, were used to excavate the modified horseshoe shaped drifts. Due to the changing orientation of the drifts as they proceed up and around the perimeter of the tunnel, the shields consisted of two parts. The inner circular shaped section contained the swing hoe excavator, the forward end of the muck conveyor, and the operator station and controls. The outer horseshoe shaped section contained the 12 100-ton push jacks. After each drift was complete, the inner section of the shield was rotated within the outer section for the next drift's orientation. The 2.9 meter diameter shield was initially 5.4 meters long, including the 1.2 meter tail skin. During the construction of Drift #1, unanticipated dry running sand was encountered. The configuration of the hoe within the shield made it impossible to excavate the face without the material running in from above the shield. Knowing that the same conditions would be encountered in Drifts 2 and 3, the top half of the leading edge of the shields were modified by adding a 0.75 meter hood. With this addition, the running sands could reach a natural angle of repose within the shield. The addition of the hood also improved the performance of the shield in the clays. Without the hood, the face of the excavation in front of the shield was bowl shaped due to the configuration of the digging hoe. In the laminated clays and silts typical in the lower drifts this shape of face tended to be unstable. The fallout associated with this instability caused overbreak in the crown of the tunnel, something to be avoided if settlement was to be minimized.

The addition of the hood gave the hoe more room to excavate and resulted in a more stable vertical face. The use of the hood was hence continued in all succeeding drifts. With the changing orientation of each drift, the hood was modified in order to maintain its position on the upper half of the shield.

The hoe excavated the material at the face and fed it onto a belt conveyor. The length of the conveyor, as typically is the case, was determined by the necessity of getting the materials car and the necessary number of muck cars beneath it. The materials car carried the precast segments necessary for one support "ring", blocks, wedges, and rail ties. The muck cars, 4 in this case, were sized to contain all material needing to be excavated to advance the shield one complete support ring, about 9 bank cubic meters.

The drift support system carried on the materials car consisted of five precast concrete segments as shown on Figure 3. The size and shape of the segments were determined by the contractor.

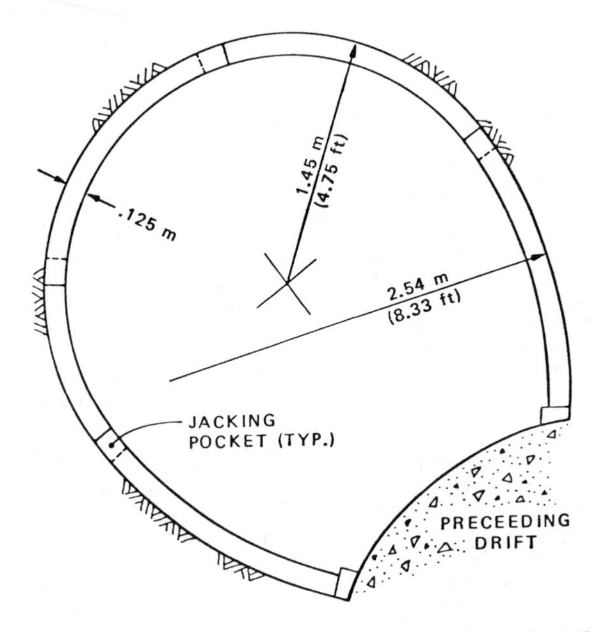

Figure 3. Cross Section Of The Precast Concrete Support System

The ruling factor in determining the thickness of the precast segments was the anticipated jacking stresses imparted to the segments by the shield jacks used to advance the shield. Anticipated ground loading stresses were checked and found to be less than the jacking stresses. The length (1.2 meters) of each support ring was determined by the size of the tunnel and the necessity of bringing the segments in beneath the shield conveyor apparatus. The length of the segments were maximized in order to lessen the number of excavation cycles needed to complete a drift.

The three segments to the shorter radius portion of the support system were not identical. We were concerned about the problems associated with riding a shield along an adjacent drift. One possible problem we felt would be the presence of a longitudinal joint at the interface. Since it was possible to eliminate this joint in the design phase, we did so. It turned out our concerns were groundless. Typically, the shield rode away from the drift and hence the presence of a longitudinal joint had no impact on shield performance. In retrospect, the three pieces should have been identical which would have reduced handling costs at the fabricator's yard and simplified stacking and erecting procedures in the drifts.

Expansion pockets cast in each segment enabled each ring to be expanded tightly against the surrounding ground with a hydraulic jack once it left the tail skin of the shield. This expansion, a contract specification stipulation, was executed using a 20-ton hydraulic jack driven by a portable hydraulic pump. Once the set had been expanded, the resulting open joint was blocked with folded softwood wedges and the hydraulic jack released. Only the joint or joints nearest springline of an individual drift needed to be jacked. Overhead longitudinal segment joints were restrained from differential radial movement by the use of H clips cut from W6 x 8.9 material. These clips proved to be an effective and inexpensive solution to the potential of a ring collapse due to differential movement of the segments before they were expanded against the ground. The open end of the horseshoe shaped support system bore against the previously constructed adjacent drift. Steamboat jacks were used to keep this portion of the support system in place until after the set had been expanded. Once the support ring had been expanded tightly against the ground, these jacks could be removed without jeopardizing the stability of the support system.

An early problem encountered during the excavation of the drifts was the control of line and grade. The design concept depended on the accurate excavation of each drift in relation to its previously constructed neighbor. The shield operators were learning to control the shield during the construction of Drift #1. This combined with the appearance of the sands within the drift horizon resulted in a significant departure of 200 mm in line and 400 mm in grade. This error was significant but was rectified by modifying the line and grade of the adjacent drifts in order to reduce the drift to drift eccentricity.

Part of the control problem was a result of the open horseshoe configuration of the support system. Since there was no support segment at the interface with the previously constructed drift there could be no shield push jacks in this area. Since the shield is "steered" by, among other things, a differential pushing with the jacks, it was easier to steer the shield toward the adjacent drift than away from it. Knowing this, and to correct minor deviations in line and grade of the preceding drift, the operators tended to keep the shield riding away from the adjacent drift. This accumulative growth in the position of the drifts would have resulted in a crown drift too narrow to excavate. To help alleviate the problem, 25 mm of the 125 mm "bottom" plate on each shield was removed. Even with the modification to the shield the crown drift was from 1.7 m to 2.1 m wide instead of the planned 2.9 m.

The other and most significant early problem was subsurface settlement occurring above the drifts. During the excavation of Drift 1, five instrumentation stations were passed. Subsurface settlement at each of these stations was less than the specified 25 mm. After excavating approximately 100 m of Drift #2 the drift drive was shut down at the direction of the Owner. At an instrumentation station passed 60 meters into Drift 2, subsurface settlement as measured 1 m above the crown, had exceeded 75 mm. This was well in excess of the specified 25 mm and was considered unacceptable given the design reliance on the undisturbed strength of the soil. After lengthy discussions between all parties concerned, the excavation was allowed to proceed with minor modifications to the support system expansion procedures. A go-slow approach was adopted and simultaneous drift excavation of Drift 3 was delayed until Drift 2 had been excavated and concreted. Drift 2 passed three additional instrumentations stations. At the first of these the settlement was again in excess of 75 mm. The settlement of the last two stations was within the specified 25 mm. Drift 3, allowed to proceed once Drift 2 had been excavated, concreted and grouted, passed six instrumentation stations. The settlement at all of these stations was within the specified limits. Based on these results Drift 4 and succeeding drifts were allowed to proceed. During the excavation of Drifts 3 and through 23, over 60 instrumentation stations were passed. Subsurface settlement at all of these, except for one over in Drift 8, were within the specified 25 mm.

An early decision was to carry out all drift excavation from the West Access Pit and to perform all drift concreting operations from the East Access Pit. This allowed excavation and concreting operations to proceed simultaneously without interfering with each other. Once a drift was holed through at the East Access Pit, the shield and trailing gear were trucked around to the West Access Pit and prepared for the succeeding adjacent drift. Meanwhile, the cleanup and concreting of the just excavated drift could commence.

The most practical method to place the backfill concrete in the drifts was to pump it from outside. The pump was rail mounted at drift level within the access pit. Concrete was fed to the pump from the edge of the pit via an inclined drop chute, a 3 cubic meter surge hopper, and a 15 m long swivel mounted traveling conveyor. The concrete is pumped to the riser car located at face through a rail mounted 150 mm diameter slickline. From the riser car to the discharge point the slickline is suspended from hangers in the arch of the drift. The end of the discharge line is kept buried from one to three meters into the sloping face of the concrete in order to force the concrete up into the arch of the drift.

In a typical tunnel, where concrete is placed between surrounding ground and the tunnel form, the flow of concrete is impeded and the slope of the advancing face of concrete is typically 3 or 4 horizontal to 1 vertical. In the backfilling of the 3 meter wide drifts, slopes of 10 to 12 horizontal to 1 vertical were common. Even with this flat slope and the distance the concrete flowed, little evidence of segregation was noted. The configuration of the placing negated the possisility of vibrating the concrete. Any vibration tended to collapse the wet concrete away from the drift arch. Rock pockets were a problem only when the concrete was placed too dry. Typically, the slump at the pump was between 125 mm (5 inches) and 175 (7 inches).

During a pour as the concrete face advances, the rail mounted riser car, slickline and concrete pump are pulled back by an air tugger anchored in the pit and connected to the pump via cables. Due to the use of the swivel mounted conveyor the concrete placement can continue as the pump is moved back. Once the pump has moved back the length of one slickline segment, 9 meters, the pump is stopped, the slickline segment nearest the pump is disconnected and swung out of the way by the service crane. The pump is then run forward by the air tugger, reconnected to the remaining slickline and pumping is resumed. This procedure takes from one to two minutes. It takes from 5 to 6 shifts to place the 2500 cubic meters per hour when pumping the full 400 meter length of the drift to just under 100 cubic meters per hour when near the portal.

After concreting, the drift is grouted to insure that any and all voids are filled. The 400 meter drift is divided into 8 equal sections for grouting. Each length is grouted through a 25 mm diameter pipe installed in the arch of the drift from the portal to the length drift it serves. Here, between each standard length of pipe, tees and nipples are installed. The nipples, plugged with a plastic cap to prevent anything entering and plugging the line, project halfway up into holes drilled through the precast liner. The injected grout is thus free to flow both inside and outside the liner. Once the reach of drift served by a particular line is concreted, it is grouted with pressures of up to 6 atmospheres. This pressure is higher than is normally used (1 to 2 atmospheres) for backfill or contact grouting but was dictated by the concern to completely fill all voids. Dye tests indicate that a fair amount of this grout is injected out into the surrounding soils along bedding planes and fracture joints. This may well have contributed to the successful minimizing of subsurface settlement.

Conclusions

Mt. Baker Ridge Tunnel has shown that large diameter soft ground tunnels can be constructed with conventional equipment by the use of the stacked drift method of construction.

The methods developed and the lessons learned should enable designers and contractors to design and construct a similar tunnel in the future with a greater sense of assurance that it can be constructed successfully.

Acknowledgements

Photographs by Washington State Department of Transportation

FLINT TUNNEL - A Project of Change and Challenge

by

Roger B. Van Omen, Vice President
Greenfield Construction Company, Inc.*

Abstract

The talk will deal with construction of a 12' diameter (8.5' finished) tunnel in the City of Flint, Michigan. This 5-mile long sewer tunnel was constructed through ground conditions varying from rock to running ground, free air to high air, low cover to high cover and, of course, varying conditions in between.

It was constructed with two different types of tunnel boring machines, a Decker-Robbins arm excavator and a Lovat wheel excavator. The project also went through a protracted and expensive arbitration proceeding over changed conditions. The problems, the solutions and the lessons learned will all be discussed.

Text

During the mid-1970's the City of Flint, Michigan set about on an ambitious pollution control program to separate its sewer system and reduce the pollution of the Flint River. During 1977 and early 1978, design and soil investigations were completed and the projects were advertised for bids in April, 1978.

The two contracts which comprise this project were a part of about a dozen contracts on the entire pollution control program. Bids were received in mid-May, 1978 and Greenfield was successful bidder on three of those contracts. Contracts 4 and 5 were adjacent tunnel contracts. Contract 9 was an open cut contract. Contracts 4 and 5 called for the construction of about 4.5 miles of 8'-6" finished diameter tunnel. The tunnel on these two contracts was the primary feeder that connected the center of the City of Flint to the treatment plant located west of the city.

The specifications for the project called for the tunnel to have a secondary lining with a finished diameter of 102 inches and a

*
Greenfield Construction Company, Inc.
13040 Merriman Road
Livonia, MI 48150

minimum wall thickness of 14 inches of unreinforced concrete. The primary liner was specified as five inch by 16 pound circular steel ribs at four foot centers with four inch hardwood lagging. Greenfield elected to mine out a 144 inch outside diameter tunnel. This would allow for a line or grade deviation of two inches in any direction and still meet the specified wall thickness requirements.

Two tunnel boring machines (TBM's) were used to construct this project. The first TBM was manufactured for the project by Carl Decker, Inc. of Detroit. It was an arm excavator type of machine that utilized a Robbins design. It became commonly referred to as the Decker-Robbins TBM. The second TBM utilized was a wheel excavator type of machine that was designed and manufactured by the Lovat Tunneling Equipment, Inc. of Toronto, Ontario - Canada.

It was our intention when the project started in the summer of 1978 to use the Decker-Robbins TBM to mine the entire 4.5 miles on both contracts. Although two separate contracts were involved, we viewed them as a single project. It was only after numerous complications and delays occurred, that it became evident it would take two TBM's to finish the project in any reasonable time. In the fall of 1980, the Lovat TBM began work on the project. Both TBM's were manufactured specifically for this project.

The Decker-Robins Back-Hoe Excavator TBM, although somewhat similar to other back-hoe excavator TBM's, incorporated a new design concept which virtually designated it as a proto-type TBM. Previous back-hoe excavators were supported off the shield of the TBM, while in this machine the excavator was supported off the conveyor frame.

Some of the machine features were as follows. The excavator assembly consisted of a traveling boom with a rotating tool for excavating the face and moving muck into the conveyor. The excavator is attached to a carriage that slides along the conveyor frame. Fore and aft movement is provided by the slide frame and a telescoping boom. Lateral and vertical movement is provided by actuating jacks. Unlimited maneuverability is provided by combining these movements. All movements are powered hydraulically. The operator is able to engage the face at any desired point including ahead of the leading edge of the shield.

The excavating tool is a combination ripper and scraper, with a ripper tool mounted opposite the scraper. The tool is rotated by a hydraulic motor. The controlled rotation capability allows the excavator operator to direct the ripping tooth at any point on the face. He can quickly change from the ripping or excavating mode to the mucking mode by rotating the tool.

In softer formations the shield accomplishes the larger share of the excavation by caving material into the center opened by the excavator. In very dense material such as hard dry clay, cemented sand or "hard pan," the ripper tooth may be used to excavate ahead of the shield cutting edge in those areas too hard for the shield to penetrate.

The excavated muck or soil is removed from the back of the apron by the machine conveyor. This conveyor is located well forward in the front of the shield under a longitudinal slot in the apron. Muck is pulled into the slot by the scraper blade of the excavator and by the crowding action of the shield and apron.

The conveyor is supported in the front shield on a flexible pivot thereby accommodating lateral movement of the rear of the conveyor when negotiating curves. The trailing end of the conveyor is supported by the power-pack structure which is attached to a skid-mounted trailing floor. This power-pack structure is so designed as to bridge the muck cars to allow their storage under the conveyor. Pumps, motors, tanks, etc. are symmetrically mounted on both sides of the frame next to the tunnel wall.

The hydraulic equipment was designed for a continuous working pressure of 3,000 psi. All valving was air controlled rather than electric. The electric horse-power furnished to the TBM was 300 h.p. The back-hoe excavator and conveyor were operated completely separate from the shield and each unit had its own power section.

The Lovat TBM was also a 144 inch outside diameter machine. It was a more conventional wheel excavator type machine, with a rotating cutter head excavating the tunnel muck. Richard Lovat builds an excellent TBM and this machine incorporated all of the latest design features in the industry and was a state-of-the-art TBM.

The wheel or cutterhead of this TBM was articulated which provides the machine operator with precise control of the TBM alignment. This feature is especially important on projects that contain tight radius curves. As a contractor, we consider any radius under 300 feet as tight, any under 200 feet as very tight and under 150 feet as not practical.

The cutterhead was powered by three 200 h.p. electric motors for a total horse-power of 600. The cutterhead is driven by six, low speed, high torque radial piston motors.

Mounted in the cutterhead are flood control doors. These doors are hydraulically operated and can completely close the face of the TBM. They give the machine operator positive control of the tunnelling face in soft or running ground conditions. The heavy plate doors also provide protection from blasting operations if rock or boulders are encountered.

The muck is removed from the face by a 25 foot long, 30 inch wide, inclined primary conveyor and a 100 foot long trailing conveyor mounted on a trailing support structure. This support structure also houses much of the support machinery and equipment necessary to run the machine.

This TBM has more power for its size than any previously owned by Greenfield. The 600 horse-power electric motors generate over a million foot pounds of rotational torque. Twelve 100 ton hydraulic propulsion jacks generate a propulsion thrust of about 2,400,000 pounds against the face.

Every bit of the power and versatility of these tunnel boring machines was needed on this project. Soil borings on the plans indicated a variety of soil conditions. The starting 700 feet of tunnel was indicated as weathered shale and sandstone with very low RQD's. The next 11,000 feet was indicated as silty sand, clayey silt, silty clay, sandy clay, sandy silt and sand and gravel. They also mentioned that cobbles and occasional boulders might be present. Another brief encounter with weathered or broken sandstone several hundred feet long was indicated at about the 12,000 foot mark. Then back into another 7,000 feet of mixed sands, silts and clays. The last 5,000 feet was constructed in primarily sands and sand and gravel. The last 2,000 feet of this 5,000 foot contained numerous cobbles, with some areas containing more cobbles than sand and gravel.

The soils we found were in some cases considerably different than indicated and in other cases the soils reacted much differently than we expected. Both of these conditions made the project much more difficult and more expensive to construct.

A preview of things to come occurred 500 feet into the project as we crossed under the Flint River. We were suddenly inundated with water coming through the crown of the tunnel. The entire tunnel, including the TBM, were under water for a week or so until we could seal the leak in the river bottom. This was accomplished by building a dike over the line of the tunnel, with a series of 60 inch culverts thru it to carry the river. After dewatering the tunnel, we discovered that the flooding had occurred thru a soil boring drilled on the tunnel alighment in the river bottom and thru the sandstone rock we were tunnelling.

Tunnelling under rivers is a hazardous undertaking anytime, without the soils boring people making it more hazardous. This is not the first time a tunnel has been flooded because a soil boring was placed on the tunnel alignment in or next to a river or body of water. Geotechnical engineers should always instruct their drillers never to drill test borings in the tunnel alignment, but especially not to drill near the tunnel alignment when bodies of water are present.

Our next surprise came once we crossed the river a few hundred feet where the borings indicated the rock would go away and we would be tunnelling in what was described as silty sand. Well, silty sand turned out to be a full face of solid sandstone. The sandstone continued for about the next 2,500 feet and had a significant impact on the amount of tunnel we were able to construct per day.

Because the conditions in the tunnel differed significantly from those indicated in the contract documents and because our costs to construct the tunnel had increased so drastically, the project was shut down in late 1979. The dispute between the contractor and the owner was submitted to an arbitration panel for settlement. The arbitration was concluded in the summer of 1980 and we went back to work immediately. By this time, the project was so far behind schedule a second TBM was necessary in order to get the project concluded in a reasonable period of time.

Once the tunnel face cleared the rock area and returned to the glacial tills indicated in the contract documents, the major problem became groundwater. The rest of the project (85%) required dewatering. This dewatering was complicated by the fact that much of the tunnel was over 60 feet deep and about 6,500 feet was 95 to 105 feet deep.

We were able to get most of the water out of the ground by the use of deep wells. In many areas this was supplemented by putting the tunnel under air. An unfortunate side effect of the dewatering was that part of the project ran through a portion of Flint Township which had no water system, only individual wells. Our dewatering system allegedly dried up their wells. This resulted in another trip to the courts to determine if the allegations were true and if they were true, who was liable for the damages. That one is still being sorted out by the lawyers.

Throughout the deepest portion of the tunnel the bedrock would periodically jump into the tunnel and then dip out of sight for awhile, only to return a little later to harass us again. About 20% of the project was ultimately constructed in either a full face or mixed face of rock.

In many areas the soil contained enough silt or clay so that the wells were not totally effective. In these areas we put air pressure on the tunnels to keep the water out of the tunnel. About two-thirds of the project was constructed under air to supplement and assist the dewatering wells. Air pressures as high as 15 psi were used, although most of the time air pressures were in the 6-9 psi range. As the cover decreased and the sand content increased, the air pressure was removed from the tunnel because we were losing too much air through the ground.

We had significant surface subsidence problems at several locations throughout the project. In some cases they were caused by the fact our dewatering system was working too well. We had removed all the water from the sand, so that the sand was now completely dry and very unstable. As our Lovat TBM wheel churned into the sand layers, the crown of the tunnel at the face of the TBM would cave in faster than the TBM could advance to support it. As a result we were taking added material into the TBM and leaving voids over the top of the tunnel.

This resulted in considerable settlement of the streets and on one occasion, at a point where the tunnel was about 95 feet deep, a 30 foot wide, 15 foot deep hole suddenly appeared in the street. The hole extended into the front lawn of an adjacent home. Needless to say, the gentleman who owned the house was concerned that it too would be swallowed up like the street, the sidewalk, the watermains and his lawn had been, so he moved out for a few days. It wasn't swallowed however, and in a few weeks the hole had been filled, the voids had been grouted, the streets and utilities restored, his lawn replaced and all was back to normal.

There was another area of the project that the ground contained so many cobbles that the TBM sounded like a giant rattle as hundreds of cobbles rattled around its wheel. These cobbles also caused so much vibration and were so unstable, that we had significant surface failures. Fortunately, this condition occurred in an area where no significant surface features were present to be damaged. At one point, as fast as the sand, gravel and cobbles were removed from the tunnel, they were trucked around and dumped back into the developing hole above the TBM. I think some of these cobbles made the trip several times that day.

The project was substantially completed in the fall of 1981, some three plus years after it began. In spite of numerous problems with the ground and several court battles with the owner, the engineer and the township, in the end the project was completed with all parties working together to achieve a common goal - an improved sewer system for Flint. Persistence and cooperation won out in the long run. It just took a little longer than we thought.

What did we learn that may be of value to future generations. I think there were several things worth noting:

1.) It is not nice and also very expensive to call sandstone, silty sand.

2.) Never drill soil borings on the tunnel alignment. Near by, but not <u>on</u> the alighment.

3.) Owners, engineers and contractors can work together successfully, but only when they realize that all the parties to the contract have equal rights.

4.) When the going gets tough - dig in. Determination, fortitude, and persistence can overcome many problems.

This presentation was supplemented by slides of the machinery and various aspects of the project.

SOFT GROUND TUNNELING OF THE GREENBELT TUNNELS

By J. Marshall Dean*

ABSTRACT

Mergentime/Loram's contract for the construction of Section E-ld Greenbelt Route Tunnels includes construction of dual 1,770-foot tunnels in soft ground, a crossover structure, a fan structure and pumping station, and other related work. This presentation describes the overall construction requirements of the Contract, including subsurface conditions; Mergentime's value engineering of the tunneling and compressed air contract requirement; selection of the tunnel boring machine; installation and use of the compressed air plant including the man, muck and medical locks; the use of precast concrete tunnel liners; and the tunneling performances during construction of the tunnels.

INTRODUCTION

Section E-ld of the Greenbelt Route, located in Northwest Washington, DC along 7th Street between H and M Streets, includes construction of dual, single track tunnels in soft ground which interface with the north end of Gallery Place Station and run 1,770 feet along 7th Street to the cut and cover crossover structure south of the Mt. Vernon/UDC Station presently under construction. Optional linings of precast concrete, cast iron or fabricated steel were specified.

Also included are a fan structure and pumping station located at 7th and K Streets, utilities, paving, and other miscellaneous work.

The joint venture of Mergentime Corporation (sponsor) and Loram Construction, Inc. was low bidder at $25,773,15 when bids were opened on April 27, 1983 by the Washington Area Transit Authority. Contract 1E0014 was awarded June 29, 1983 with Notice to Proceed August 19, 1983. Time of completion for the contract was 1,080 calendar days.

*Assistant Vice President, Tunnel Division, Mergentime Corporation, P.O. Box 662, Flemington, New Jersey 08822

The Special Provisions of the Specifications required the following:

Tunneling was to be started at the north end of the Contract from the cut and cover section and to proceed southward to Gallery Place Station with the outbound tunnel first, followed by the inbound tunnel. The tunnel shields were to be abandoned at Gallery Place Station. Ground control was to be achieved using either compressed air or a pressurized face shield alone or in conjunction with partial dewatering. Free air tunneling was allowed for the first 150 feet of tunnels using chemical grout as ground control. Under the compressed air option, a minimum of 12 psig air pressure was to be maintained in the tunnels at all times. The air-compressing plant for the driving of each tunnel under full air with no dewatering was to be capable of furnishing a minimum of 9,000 cubic feet of free air at a pressure of 35 pounds per square inch above atmospheric pressure at the discharge point of the compressors. At least 1,800 cubic feet of outside air per man per hour was to be provided in the working areas under pressure. A minimum of one man lock and one materials lock were to be provided in each tunnel. A medical lock was to be provided on site. Compression and decompression of persons entering and leaving areas under compressed air were to be undertaken in accordance with applicable federal regulations including OSHA.

Specified buildings on the east side of 7th Street were to be protected by compaction grouting. Compaction grouting, a relatively new technique used to protect buildings adjacent to tunnels under construction, involves the injection of low sump grout under high pressures which will fill up the soil voids created by tunneling before settlement occurs to the adjacent buildings.

An instrumentation program specified that the Contractor monitor ground water levels, surface and subsurface earth settlements, and surface structure settlements. The program was to provide reliable, early information to permit timely action to minimize or prevent damage to structures and utilities.

GEOLOGY

The Geotechnical Basis of Design and of Construction Specifications Report was included in the specifications. The information contained in the report included the subsurface investigation reports by the General Soil Consultant, Mueser, Rutledge, Johnston and DeSimone, formerly Mueser, Rutledge, Wentworth and Johnston.

The geology along Section E-1d is characterized by the Cretaceous soils of the Potomac Group overlain by the terrace deposits of younger Pleistocene Age. These soils were deposited by rivers during periods of erosion following regional uplift. The bedrock is schistose gneiss overlain by decomposed rock, the top of which is located several tunnel diameters below the tunnel invert. The Cretaceous deposits and Pleistocene terrace deposits have been subdivided by the General Soil Consultant (GSC) into generalized strata descriptions.

The Potomac Group of the Cretaceous deposits are designated P1 to P4. The Pleistocene terrace deposits are designated T1 to T5. The GSC describes these strata as follows:

P1- Hard plastic clay with occassional pockets of fine sand.
P2- Compact to very compact silty or clayey fine to medium sand with traces of small gravel.
P3- Hard silty or sandy clay and silty or clayey fine sand with some small gravel.
P4- Very compact silty or clayey fine to medium sand with some gravel and boulders.

T1- Stiff to medium silty clay or clayey silt with fine sand lenses.
T2- Medium compact to compact silty or clayey fine to medium sand with trace of small gravel.
T3- Compact to very compact fine to coarse sand with some silt and gravel and numerous boulders.
T4- Medium compact to compact fine to medium silty sand and small gravel with lenses of dark clay.
T5- Compact to very compact fine to coarse silty sand with some gravel and numerous boulders.

Tunnel excavation will lie generally within the P1, P2, and P3 strata. Contact between the Cretaceous P series and the overlying Pleistocene T series occurs within four-five feet of the crown for the first 400 linear feet of the tunnels and penetrates as far as seven feet below the crown for approximately 400 linear feet at a distance of 1,165 feet along the alignment.

The groundwater table ranges from 12 feet to 30 feet, averaging 20 feet for the most part and generally rising in the northerly direction. Thus, the height of the groundwater table above the tunnel invert was anticipated to range from a minimum of 30 feet at the northern limit, increasing to 50 feet, and then gradually decreasing to 40 feet at the southern limit.

VALUE ENGINEERING

Prior to the award of Contract, Mergentime retained the services of a Board of Consultants. This board consisted of Dr. Edward J. Cording, University of Illinois; Mr. John P. Nunan, P. Eng., Vice President and Principal Hydrogeologist, Hydrology Consultants, Rexdale, Ontario; and Robert G. Lenz, Chairman, Moretrench American Corporation and President, Ground/Water Technology, Rockaway, New Jersey.

Their initial task was to undertake a test dewatering program and investigation. The purpose of the investigation was to ascertain whether by proper dewatering the air pressure in the tunnels could be reduced, and whether dewatering wells could be used in lieu of chemical grouting at the start of the tunnels.

Our Consultants stated that favorable results could be obtained by dewatering the reaches of tunnel and arrived at the following conclusions:

1. Free air tunnelling for 300 lf was feasible.
2. Dewatering was recommended as an alternate to the chemical grouting of the first 166 lf of tunnel.
3. Air pressure should not be specified as a minimum of 12 psi since "minimum air pressures of 12 psi will be too high in many sections of the tunnel."
4. Dewatering can result in using air pressures of five psi or less.
5. A 35 psi air compressor plant was not required.

Based on the above conclusions, we proposed to the Authority a Value Engineering Change Proposal (VECP) for the following changes to the specifications:

1. Free air tunnelling for the first 300 lf of tunnel in lieu of the first 150 lf of tunnel.

2. Dewatering as an alternate to the chemical grouting of the first 166 lf of tunnel.

3. Elimination of the 12 psi minimum air pressure.

4. The installation of a 15 psi air compressor plant in lieu of a 35 psi air compressor plant.

5. Reversal of the tunnel driving sequence, starting the inbound tunnel first.

The use of the Lovat Tunnel Boring Machine and our dewatering program was a major factor in evaluating our VECP. The Lovat Machine had capabilities which would make the task of controlling ground water and air pressure easier than with other machines, and it therefore represented a capability of dealing with potential difficulties in a more effective way. In addition, we were instituting a dewatering program which would lower the water table approximately to the crown of the tunnel.

Mergentime presented the VECP to the Authority. After a complete review by the Authority's staff and consultants, The Contracting Officer responded favorably to the minimum 12 psi air pressure requirement being reduced except in the areas of those buildings which were to be protected by compaction grouting; and approval of a new minimum compressed air plant capacity. Substantially disagreeing with the response, we requested a meeting with the Authority's Board of Consultants so additional presentations could be made. As a result of this meeting at which Mergentime proposed an additional extensive monitoring system program, we received approval of all the proposed changes following an acceptable agreement for cost savings.

After successfully driving 300 linear feet of tunnel under free air we proposed a second VECP to drive tunnel without the use of compressed air beyond the 300 foot point until such time as conditions required compressed air. This proposal was also accepted following agreement for cost savings.

SELECTION OF EQUIPMENT

The first task was to finalize the selection of the Tunnel Boring Machine. Extensive studies were undertaken of all the tunnel machines offered. Top management traveled to Germany, Italy, and Brazil to see soft ground TBMs working. After thorough evaluations of all the machines, a Lovat Tunnel Boring Machine was selected in August 1983, and a contract awarded to Lovat Tunnel Equipment-USA September 23 for the manufacture and delivery within 7.5 months of a Model M225 Tunnel Boring Machine and ancillary equipment. This machine, 18 ft.-10 in. in diameter by 20 ft. nominal in length would be designed to be dismantled and removed from inside the shell and for reassembly within a new shell. It would have an articulated cutting head equipped with hydraulically operated flood control doors for positive face control in soft ground and mixed face tunneling conditions and have a heavy duty cone, hub and spoke assembly. The cutting head rotation would be 0-4 RPM at variable speed, with clockwise and counterclockwise rotation. The cutting head was to be powered by four-200 hp electric motors driven by ten hydraulic motors. The type of teeth would be determined after analysis of soil reports and samples. The M-225 was to be propelled by 18-150 ton push jacks with 68 in. stroke and operated by a 6,000 psi hydraulic system powered by a 100 hp electric motor. The push jacks would bear against a pushing ring. The machine would be equipped with an erector for placing precast concrete lining segments into position. Other features included four stabilizer fins to anchor the machine against rotational forces, an automatic tilt control and a gas detector with automatic power shut off control. The trailing shield would be fitted with a grout seal to prevent grout from entering the shield. All electrical components would be explosion proof. Ancillary equipment included a 25 ft. long, 48 in. primary belt conveyor mounted and supported within the main shell with two hydraulically operated augers mounted to facilitate the removal of excavated material. The primary conveyor would be hydraulically retractable from the cutter head to permit access to the wheel face. The 48 in. trailing belt conveyor, 100ft. long, would be made up of detachable and interchangeable sections and be post supported and mounted on pipe skates. Each conveyor was to be driven by two variable speed hydraulic motors.

The selection of the tunnel bulkhead and lock equipment for the compressed air operations followed. The bulkhead and tunnel locks were to be designed to a maximum working pressure of 35 psig.

Prior to completing the requirements for the sizes and lengths of locks, the tunnel haulage system had to be selected. It was decided to use a 36 in. gauge, two train system with one battery locomotive operating within the compressed air tunnel and one battery locomotive operating in free air. The first train would be made up of two concrete segment cars (three segments/four segments), one 70 cf grout car, and four eight cy muck cars. The second train: four eight cy muck cars. The maximum length then was determined to be 90 ft. This allowed selection of the length of muck lock at 100 ft. The design and fabrication of the bulkhead, tunnel locks, and medical lock was awarded to Elgood Mayo Corporation, Lancaster, PA. The bulkhead design was a basic four piece steel bulkhead supported on an imbedded anchor ring. The manlock was 5 ft.-6 in. in diameter by 22 ft. long with the mucklock 9 ft.-8 in. in diameter by 100 ft. long. The medical lock was an OSHA (75 psig) design, 6 ft.-6 in. in diameter by 20 ft. long with two compartments.

Locomotives selected were 20 ton, two motor drive storage battery locomotives. Lead acid storage batteries were 64 cell, 25 plate, 1,440 amp-hr.

For grouting behind the concrete tunnel lining the following equipment was selected:

Grout cars - 70 cf with mixer/agitator, and 6 in. Moyno pump, all air driven.

Grout pump and hopper assembly - 70 cf holding tank with agitator and 6 in. grout pump all hydraulically driven. This assembly was mounted on the primary conveyor close to the shield of the TBM.

Grout plant - A custom designed grout plant consisting of one 85 cy sand silo; one 230 barrel flyash silo; one 50 cf bentonite hopper; 14 in. x 21 ft. long swivel auger with chute; electronic synchronized controls for product mix design control. This unit was installed in the shaft (crossover area).

The design and furnishing of the compressed air system was awarded to Compressor Engineering, Detroit, Michigan. This plant included five prefabricated piped and wired low air units. These skid mounted units were made up of Gardner Denver Cycle Blowers, Models 9CDL23 at 2100 cfm each with pressure to 18 psig, driven by 440v 200 hp motors; inlet and discharge silencers and water-cooled aftercoolers. The water recirculating system included a cooling tower and pumping station. Also included in the compressed air system was a control air system, and a master control and alarm panel. Following installation of the system a 750 kw generator was installed for standby power to the air plant.

PRECAST CONCRETE LINING

Mergentime selected the precast concrete lining as their option of liners. Again, an intensive study was undertaken to evaluate the proposals of the potential suppliers. Following a final analysis which included cost, experience, anticipated performance and delivery, Buchan Concrete Tunnel Segments, Ltd., was selected as the supplier.

Buchan chose Baltimore, Maryland as the site for their precast plant. Production casting of the segments began in June, 1984 working one shift. By July, three shifts were producing seven rings per day. Casting of all rings was complete in early December, 1984.

Buchan achieved excellent results in the precasting, handling, storing and shipping of the liners.

DEWATERING PROGRAM

The dewatering program consisted of dewatering wells installed along the tunnel alignment to provide a system which would provide stable soil conditions for either free air or low air pressure (five psi) tunneling. The number and spacing of dewatering wells varied along the route due to changes in hydrogeology and required drawdown. Generally, required drawdown for a maintained water level at tunnel crown was 20 feet but as little as nine feet. A total of 22 wells including the four test wells were installed with some wells having a vacuum assist.

MONITORING PROGRAM

The monitoring program was under the direction of Dr. Edward Cording, Geotechnical Consultant. Control to minimize ground movements and damage to structures was to be accomplished by monitoring ground movement, building settlements, water levels, and ground conditions; and coordinating these observations with the tunneling by applying appropriate levels of compaction grouting and compressed air.

Results of the monitoring program for the inbound tunnel indicated that the TBM performance was excellent, that grouting of the liners resulted in almost complete filling of the tail void, and that dewatering of the soils was effective. Settlement monitoring indicated that only small movements were developing at the tail of the shield with center line surface settlements ranging between 0.25 inches and 0.70 inches and surface settlements 20 feet off centerline ranging from 0.10 inches to 0.40 inches.

TUNNELING OPERATIONS

Prior to shipment from Toronto, the TBM was disassembled under simulated tunnel procedures to confirm the design of dismantling and removing the TBM from inside the shell. Seventeen trailer loads were required to ship the TBM and ancillary equipment to the jobsite.

Prior to shipment of the equipment, excavation of the crossover was completed. This crossover area was to be used as the shaft for tunnel operations.

Mining of the inbound tunnel in free air began on July 26, 1984 on a two shift basis. On August 7, mining went on a three shift, five day basis. Tunneling continued in free air until 300 feet of tunnel was mined. At this time, August 20, the compressed air bulkhead and locks were installed. With the acceptance of continued free air tunneling, mining was resumed on September 6. To demonstrate the capability of compressed air operations, the tunnel was pressurized to five psig on August 14. Having satisfactorily demonstrated compressed air operations for one shift, free air tunneling was resumed.

Tunneling progressed satisfactorily without air for approximately 1,425 feet. At that point, due to unstable soil and water conditions in the face, mining commenced under compressed air at five psi and continued for 217 feet. With improved soil conditions, the compressed air was turned off and the remaining 128 feet was mined in free air. The TBM "holed thru" into Gallery Place Station on November 19, 1984. The TBM was then dismantled from the shell and removed through the completed tunnel into the shaft, leaving the shell in place in the inbound tunnel.

The TBM was reassembled in the shaft with a new shell for the second tunnel (outbound) drive. An additional 40 feet of secondary trailing conveyor was added to allow the two train system to work in tandem with the segment-grout-muck train in first.

Mining of the outbound tunnel in free air commenced on January 24, 1985 on a two shift basis. On February 4, 1985 mining went on a three shift, five day basis.

The bulkhead and locks were not installed for the 300 foot free air limit. Based on the experience gained from the inbound tunnel drive, it was agreed that pretunnel consolidation grouting and post tunnel regrouting from the street surface would be more effective for ground control than compressed air.

The outbound tunnel holed through into Gallery Place Station on May 1, 1985. Tunneling had progressed satisfactorily without air for the entire drive. Pretunnel grouting and post tunnel grouting was used for ground control for the last 550 feet of the drive. Compaction grouting was also used to limit ground movements and to protect the designated adjacent structures from damage.

Overall mining progress for the inbound tunnel was 30 feet per day with best days of 48 feet. The outbound tunnel progress was 33 feet per day with best days of 48-60 feet.

CONCLUSIONS

The successful tunneling of the Greenbelt Tunnels can be attributed to the following:

1. Effective project and home office planning and management.
2. The selection of and the subsequent performance of the Lovat TBM.
3. The acceptance of the Value Engineering Change Proposals.
4. Good performance in precasting tunnel liners.
5. An effective dewatering program.
6. The planning and implementation of the Monitoring Program.
7. A good safety program and record.

CHICAGO'S TARP CHALLENGE - 8 MILLION TONS OF ROCK

by William C. Paschen*

and

Daniel F. Meyer

INTRODUCTION

The challenge involved with moving millions of tons of rock was addressed by nationally recognized tunnel contractors who had converged in Chicago over the past 5 years. This wealth of expertise was needed to meet the risk and objective of constructing the world's largest network of underground tunnels which in turn required a dramatic advance in the art and science of tunneling. These pioneers, through innovation, adaptation and hard work eliminated the one major concern of The Metropolitan Sanitary District of Greater Chicago - how to move 8 million tons of rock.

The Metropolitan Sanitary District of Greater Chicago (MSDGC) serves 122 municipalities covering 866 square miles in Cook County, Illinois. Of these cities, towns and villages, 53 have combined sewer systems. On an average of once every 4 days, these systems overflow about 4 million gallons of combined sewage and rainwater into the area's rivers and canals. The overflows occur at approximately 640 outfall locations of which 10% are owned by the MSDGC, with the remainder of the points belonging to various other municipal bodies. All those who own these "point" sources of pollution are in violation of Illinois State and Federal regulations.

To enable its member municipalities to come into compliance with State and Federal law, the MSDGC is implementing its Tunnel and Reservoir Plan (TARP). TARP is a regional solution to the problem of combined sewer overflow in the metropolitan Chicago area.

TARP will greatly reduce the pollution of area waterways by intercepting the combined sewage and rainwater carried in the existing systems and diverting that flow into deep rock tunnels. The tunnels will carry the flow to reservoirs, where it will be held until it can be pumped to sewage treatment plants for purification and final

* Vice-President, Paschen Contractors, Inc., 2739 North Elston Avenue, Chicago, Illinois 60647

discharge into the waterways. When completed, the program also will greatly reduce flooding in the combined sewer areas.

TARP is divided into 2 parts: Phase I for pollution control and Phase II for flood control (reference Figure 1). Phase I consists of about 110 miles of rock tunnel, 252 drop or inlet shafts and approximately the same number of connecting structures. Phase I is estimated to cost $2.2 billion with 75% grant funding from the United States Environmental Protection Agency. $1.2 billion in contracts have already been awarded and Congress gave responsibility for engineering and implementing Phase II to the United States Army Corps of Engineers in 1976. Phase II consists of about 21 miles of tunnel and 3 reservoirs. The most recent cost estimate for Phase II is $800 million.

The tunnels are founded in dolomitic limestone at a depth of 150'-300' below ground surface. Currently, 47 miles of tunnel have been completed. Of the 4 subsystems of TARP, the 6.6 mile O'Hare System has been completed, 31.2 miles of the 40.3 mile Mainstream System and 9.2 miles of the 36.3 mile Calumet System. No work has been started on the 26.4 mile Des Plaines System. Dedication ceremonies for the Mainstream System were held on May 23, 1985. Future construction of Phase II awaits Federal study and funding, which is anticipated.

TUNNEL BORING MACHINE (TBM) UTILIZATION

Bedrock in the Chicago area is dolomitic limestone with unconfirmed compressive strength ranging from about 12,000 psi to 25,000 psi. The rock mass is generally homogeneous with very few significant fault areas. Although water is present in the rock, the larger quantities generally appear at the various interface lines such as at the over-burden-rock line. Some shale beddings are significant in thickness.

Due to the generally favorable geologic conditions, the use of TBM's or Moles was specified by the MSDGC for all major tunnel excavations, with only short tunnels and irregularly shaped structures permitted to be excavated by conventional blasting methods. Ten mining machines varying in size from 35 ft 4 in to 14 ft 2 in OD were used on Phase I construction with 6 orders going to The Robbins Company and 4 orders going to Jarva/Atlas Copco.

Table I lists significant characteristics of those machines.

TRANSFER & HAULAGE EQUIPMENT

Although much of the transfer and haulage equipment generally were similar on the various projects, the variances in tunnel diameter and length required various combinations of that equipment (reference Table II and III).

All contractors utilized an trailing floor or mobile gantry consisting of a number of individual rolling platforms (approximately 20' in length) coupled into a continuous mobile platform ranging

FIGURE 1

TUNNEL AND RESERVOIR PLAN
Metropolitan Sanitary District of Greater Chicago

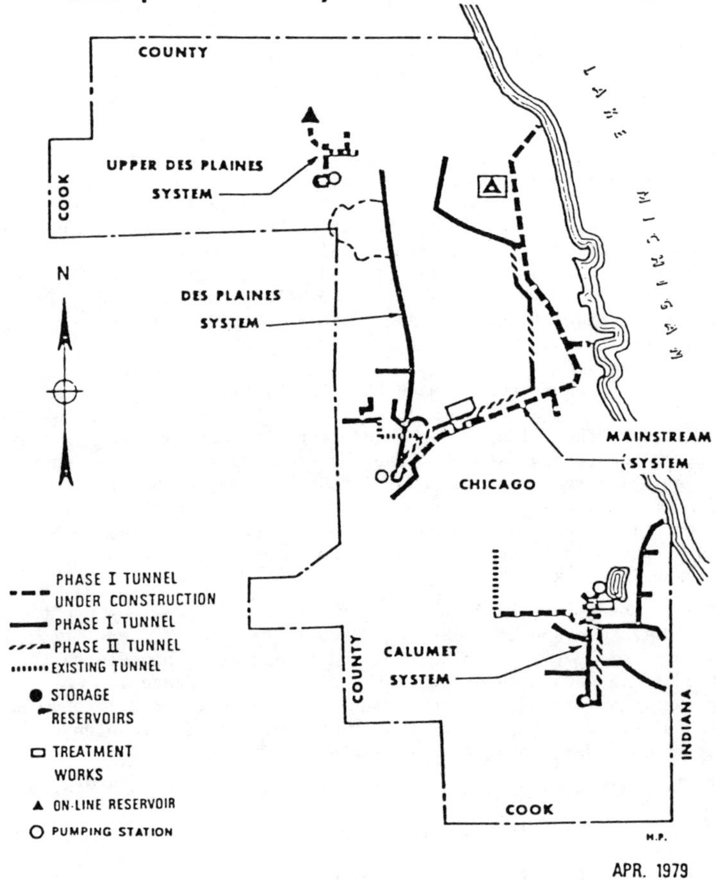

TABLE I

CONTRACTOR	MACHINE CHARACTERISTICS (ROBBINS)						
	OUTSIDE DIAMETER	CUTTERHEAD HORSEPOWER	CUTTERHEAD THRUST	CUTTER			OTHER
				NUMBER	SPACING	DIAMETER	
Morrison-Knudsen, Kenny, Paschen, S & M	35ft 4in	2600	2,760,000 lb	69	3.50 in	15.5 in	Roof bolt station
Healy-Ball-Horn	35ft 3in	2400	2,760,000 lb	69	3.00 in	15.5 in	Roof bolt station Ring steel erector Shotcrete station
Ball-Healy-Horn	32ft 4in	2400	2,572,000 lb	64	3.00 in	15.5 in	Roof bolt station
Shea-Kiewit	32ft 4in	2400	1,950,000 lb	64	3.50 in	15.5 in	Roof bolt station Ring steel erector Shotcrete station
Traylor-Ferrera-Resco	21ft 0in	1600	1,410,000 lb	47	3.00 in	15.5 in	None
Healy-Ball-Grow	14ft 2in	900	1,240,000 lb	31	3.75 in	15.5 in	None
	MACHINE CHARACTERISTICS (JARVA/ATLAS COPCO)						
Paschen, Morrison-Knudsen, Kenny	32ft 3in	2400	3,000,000 lb	54	4.00 in	15.5 in	Ring steel erector Shotcrete station
Kenny-Paschen-S & M	30ft 1in 22ft 1in	2400 1200	3,000,000 lb 2,000,000 lb	65 41	3.00 in 3.50 in	15.5 in 15.5 in	None
Shea-Kiewit	15ft 3in	750	1,200,000 lb	28	3.50 in	15.5 in	Shotcrete station

in length from 265' to 780'. The trailing floor or gantry was towed with cables attached to the rear of the TBM and the module rode on tunnel track laid directly behind the mining machines.

Empty muck trains moved up a ramp onto the trailing floor, loaded with tunnel muck and returned to a work shaft for transfer of the rock to the surface. The trailing floor was also the superstructure base which supported conveyors, electrical transformers and switch gear, ventilation pipe erectors and other required equipment.

The 3 types of trailing floor based loading devices used on TARP were the car passer, moveable tripper and train-under-conveyor arrangement.

Car Passer

Four contractors used a hydraulic car passer system. During a typical loading cycle, an empty train approached the floor mounted car passer on the inbound track located on the right side of the trailing floor. The lead car was automatically uncoupled and positioned on the car passer, which was a chain driven platform that moved the car horizontally across the width of the floor and aligned the car with the outbound track. An automatic car pusher moved the car toward the loading point and coupled it with an existing string of loaded cars. The car was then gradually passed under the loading conveyor hopper where it was charged. As each car was loaded and cleared the car hopper, the mechanism returned to the inbound track side and the cycle was repeated.

With this system the only conveyor utilized (excluding the conveyor within the TBM structure) was a bridge conveyor from the rear of the TBM to the fixed location loading hopper.

Moveable Tripper

Two contractors chose a moveable tripper system. This system consisted of a conveyor which was as long as a train (usually 6 to 8 cars) and it was located directly between and over the two tracks on the trailing floor deck. As material was transferred from the TBM bridge conveyor to the trailing floor conveyor, a mobile tripper or deflector directed the rock off the appropriate side of the conveyor belt and individually loaded each car. As one train was being loaded on one side of the floor, a second empty train could be placed on the opposite track in preparation for loading. After the first train was loaded, the tripper returned to its original starting position at the leading end of the floor and began loading the second train.

Train-Under-Conveyor

A third system selected by 2 contractors was the train-under-conveyor loading method. This system simply allowed for a train to be gradually moved under the end of an overhead loading conveyor. The train road a single track in the middle of the trailing floor and

TABLE II

CONTRACTOR	TRAILING FLOOR/TRAVELING GANTRY CHARACTERISTICS																					BRIDGE CONVEYOR (TBM TO FLOOR OR GANTRY)		
		UNITS					CONVEYOR(S)						LOADING MECHANISM					FANLINE ERECTION						
	TOTAL LENGTH (FT)	NUMBER	AVERAGE LENGTH (FT)	WIDTH (FT)	TRACK GAUGE (IN)	NUMBER OF TRACKS ON FLOOR	NUMBER	WIDTH (IN)	SPEED (FPM)	HORSEPOWER	NUMBER OF MUCK CARS ON FLOOR	MUCK CAR SIZE CUBIC YARDS	CAR PASSER	TRAIN UNDER CONVEYOR	MOVEABLE CONVEYOR TRIPPER	FIXED LOCATION LOADING	FLOATING LOCATION LOADING	AUTOMATIC	MANUAL	DUCT SIZE (IN)	WIDTH (IN)	SPEED (FPM)	HORSEPOWER
Morrison-Knudsen, Kenny, Paschen, S & M	400	19	15	11	42	2	-0-	---	---	---	8	25	Yes	---	---	Yes	---	Yes	---	54	42	400	20
Healy-Ball-Horn	320	7	30	25	42	-0-	2	42	450	40	8	24	---	Yes	---	Yes	---	---	Yes	60	42	276	400
Ball-Healy-Horn	360	18	15	11	42	2	-0-	---	---	---	8	25	Yes	---	---	Yes	---	---	Yes	54	42	400	20
Shea-Kiewit	258	16	15	10	36	-0-	2	36	560	25	5	25	---	---	Yes	---	Yes	Yes	---	54	42	276	45
Traylor-Ferrera-Resco	265	13	20	11	42	2	-0-	---	---	---	6	18	Yes	---	---	Yes	---	Yes	---	48	36	250	10
Healy-Ball-Grow	780	46	15	8	24	1	1	24	450	20	8	8	---	Yes	---	Yes	---	Yes	---	30	24	400	15
Paschen, Morrison-Knudsen, Kenny	381	18	20	16	42	2	-0-	---	---	---	6	25	Yes	---	---	Yes	---	Yes	---	54	54	315	40
Kenny-Paschen-S & M	290	17	15	11	36	2	1	36	400	30	6	26	---	---	Yes	---	Yes	Yes	---	54	36	350	10

TABLE III

CONTRACTOR	HAULAGE EQUIPMENT						
	MUCK CAR SIZE	LOCOMOTIVE SIZE	TRACK WEIGHT	TRACK GAUGE	TIE SIZE (TIMBER)	TIE SPACING	COMMENTS
Morrison-Knudsen, Kenny, Paschen, S & M	25 cu yd	35 Ton	85 lb	42 in	6in x 8in x 6ft 0in	24 in	
Healy-Ball-Horn	24 cu yd	35 Ton	85 lb	42 in	8in x 8in x 13ft 0in	24 in	
Paschen, Morrison-Knudsen, Kenny	25 cu yd	25 Ton	85 lb	42 in	6in x 8in x 6ft 9in	24 in	
Shea-Kiewit	25 cu yd	40 Ton	85 lb	36 in	8in x 8in x 4ft 2in	20 in	
Ball-Healy-Horn	25 cu yd	35 Ton	80 lb	42 in	6in x 8in x 6ft 0in	18 in	
Kenny-Paschen-S & M	25 cu yd	25 Ton	85 lb	42 in	6in x 8in x 6ft 9in	24 in	
Healy-Ball-Grow	8 cu yd	12 Ton	56 to 75 lb	24 in	6in x 8in x 4ft 8in	32 in	
Traylor-Ferrera-Resco	18 cu yd	20 Ton	85 lb	42 in	None	---	No ties - Rail fastened directly to rock

each car was loaded as the locomotive slowly moved the train. Inbound and outbound trains passed at remote locations since only one track was available on the trailing floor.

SHAFT BASED HANDLING EQUIPMENT

A considerable amount of dissimilarity was evidenced in the selection of shaft based material handling systems. Of the Joint Ventures listed in the previous tables, 2 selected the combination of hoist and balanced skips, 3 utilized hoists with lift off boxes, one selected a crawler crane and lift off boxes, one chose a vertical conveyor belt and one utilized an inclined conveyor belt system.

Table IV lists pertinent data for the above mentioned rock handling facilities.

Hoist and Balanced Skips

The 2 contractors who chose the hoist and balanced skip method of raising tunnel rock utilized almost identical systems. Twenty-five cu yd muck cars were discharged via a rotary car dumper into a surge bin located immediately below the dumper. Surge bin capacity was typically 125 cu yd (live) with 2 chute openings located in the bin bottom to transfer the material to measuring pockets.

Radial gates controlled the flow of material from each of the surge bin chutes to the measuring pockets. Both surge bin chutes and measuring pockets were lined with T-1 steel to preserve life. Although the pockets had a nominal volumetric capacity of 15.0 cu yd, the capacity could be adjusted upward to 16.5 cu yd and downward to 13.0 cu yd by use of a hand operated baffle plate.

The lower radial gates positioned at the bottom of the measuring pockets were interlocked with the upper gates to prevent a "flow through" condition with respect to skip loading. The skips utilize sheet shaft guides with appropriate safety dogs.

Both contractors utilized crown mounted hoists which could be operated manually, semi-automatically or fully automatically. The headframe crown was typically 80' above the ground surface. As the skip approached the upper headframe area, it entered a scroll plate and swung away from the shaft centerline and bottom dumps.

Both contractors constructed a surface bulkhead adjacent to the headframe which enabled a 400 - 500 cu yd rock pile to accumulate at the foot of the headframe. From this point, trucks were loaded with rubber tired front endloaders.

Hoist and Lift Off Boxes

Although the 3 contractors who utilized the hoist and lift off box combination had systems which were similar in gross characteristics, hoisting capacities varied significantly. Two of the systems serviced large TBM's (35' 3" OD and 32' 4" OD) and one serviced a mining machine of much smaller size (14'2" OD). Accordingly, the larger system had an overall rated capacity of

TABLE IV

CONTRACTOR	CHARACTERISTICS OF ROCK HANDLING FACILITIES															
				LINE		LIFT CAPACITY		SHAFT BOTTOM EQUIPMENT								
	HOIST (HP)	CONVEYOR(S)	LIFT CRANE	SPEED (FPM)	PULL (LB)	SKIPS (CU YD)	BOXES (CU YD)	ROTARY CAR DUMP (HP)	CRUSHER	SHAKER (HP)	SURGE BIN (CU YD)	TRANSFER CONVEYOR	SHAFT TOP EQUIPMENT	RATED SYSTEM CAPACITY CU YD PER HOUR	VERTICAL DISTANCE TUNNEL INVERT TO GROUND SURFACE (FT)
Morrison-Knudsen, Kenny, Paschen, S & M	600	---	---	285	85,000	2 at 16.5	---	40	---	---	125	---	7 cu yd Loader	460	278
Healy-Ball-Horn	600	---	---	410	91,000	---	24	---	---	---	---	---	5 cu yd Loader	550	260
Ball-Healy-Horn	600	---	---	250	85,000	2 at 16.5	---	40	---	---	125	---	7 cu yd Loader	500	255
Shea-Kiewit	500	---	---	820	70,000	---	25	---	---	---	---	---	7 cu yd Loader	500	260
Traylor-Ferrera-Resco	---	---	150 Ton	340	60,000	---	18	---	---	---	---	---	5 cu yd Loader	300	297
Healy-Ball-Grow	500	---	---	1230	14,000	---	8	---	---	---	---	---	4 cu yd Loader	200	220
Paschen, Morrison-Knudsen, Kenny	---	400 HP 450 FPM 32 in wide	---	453	20,000	---	---	30	75 HP 100 tph	50	50	75 HP 36 in	Radial Stacker 7 cu yd Loader	900	265
Kenny-Paschen-S & M	---	500 HP 750 FPM 36 in wide	---	---	---	---	---	40	125 HP 100 tph	50	100	20 HP 36 in 750 FPM	Horz. Stacker 20 HP 36 in 750 FPM	750	230

500 - 550 cu yd per hour while the smaller has a rated capacity of 200 cu yd per hour.

Of the 3 systems under discussion, one involved a crown mounted hoist while 2 involved conventional, ground positioned hoist houses.

All systems utilized list off boxes (chassis mounted) which were singularly positioned in a hoisting cartridge by a transport locomotive. The cartridge was then hoisted along shaft guides in a manner similar to the skips, except that the mechanism was counter balanced with dead counterweights.

As the cartridge approached the upper portion of the headframe, it entered a scroll plate which rotated the box away from the shaft centerline and into an inverted position at which point it discharged onto a bulkhead retained rock pile at the foot of the headframe. Muck pile capacities were in the 400 - 500 cu yd range. Rubber tired loading equipment charged trucks for ultimate material disposal.

Crawler Crane and Lift Off Boxes

A sixth system was utilized employed a crawler crane positioned at the shaft top. The machine hoisted 18 cu yd lift off boxes via a conventional, guided bridal arrangement. The boxes were dumped at the shaft top with a whip line attached to the box bottom.

The system serviced a 21'0" in OD TBM and hoisting capacity were rated at 300 cu yd per hour. Trucks were charged with conventional rubber tired loaders.

Vertical Conveyor

A seventh system employed on TARP utilized a vertical shaft conveyor belt.

In this system 25 cu yd rock cars were singularly positioned in a rotary dumper by a transport locomotive. The dumper discharged into a 50 cu yd surge bin located immediately below. The rock was removed horizontally with a vibrating shaker to a small hopper feeding a variable speed, inclined conveyor which moved the material from the car dumper area to the shaft bottom where the vertical shaft conveyor was positioned.

The shaft conveyor actually contained a short distance of horizontal run at both the bottom and top of the shaft, with the section of conveyor in the shaft being truly vertical. The belt was a continuous loop throughout all horizontal and vertical runs.

Material entered the pockets of the shaft conveyor at the lower horizontal section of the conveyor at a controlled rate and speed with the belt vertically upward through the shaft and then again horizontally at the shaft top.

The rock was discharged at the location of the upper head pulley onto a radial stacker. The stacker serviced a muck pile of about 7000 cu yd. Overal system capacity was rated at 900 cu yd per hour.

Inclined Conveyor

An eight system which was used involved an inclined conveyor system for transporting tunnel rock from the shaft bottom to the ground surface.

The operational characteristics at the shaft bottom were similar to the vertical conveyor system with a rotary dumper, shaker and short transfer conveyor.

However, the short transfer conveyor charged a long continuous inclined conveyor placed at 20° with the horizontal. This inclined conveyor was located within a slope excavation and it was through this excavation that the material was transported to the ground surface.

A horizontal stacker was used to stockpile material at the surface.

SUMMARY

Since 1975, the heavy construction industry has taken a quantum leap in the state of the art of design and use of rapid excavation and materials handling equipment in tunnels.

A new generation of tunnel boring machines in the 35' diameter range have been successfully utilized, whereas only several years ago the maximum economical TBM size was in the 25' range. This represents a 300% increase in face size.

Auxiliary excavation equipment such as the trailing floor have also been significantly streamlined with the introduction of the floor mounted car passer.

Of equal import, traditional mine equipment has been introduced into civil construction via the use of headframes and hoists. The crown mounted hoist was a unique development in Chicago.

TARP has also seen the innovation application of conventional conveyors and America's introduction to the vertical conveyor with respect to shaft based materials handling systems.

Chicago and the MSDGC's TARP problem of excavating and handling millions of tons of rock has been solved.

SUBJECT INDEX

Page number refers to first page of paper.

AUTHOR INDEX

Page number refers to first page of paper.